© 2022 Dan the Fish Publishing

ISBN: 978-1-7352026-3-1

Contact us at:

danthefishbooks@gmail.com

danthefishpublishing.weebly.com

The Miracle League of the South Hills presents
The Miracle League All Stars in

# It's Game Day!

illustrated by Chelsea Rae Meeker

It's game day at the Miracle League of the South Hills. The staff, coaches, and volunteers are getting the field ready to go.

Buckets full of balls and assorted colored bats are out, ready for players to arrive and warm up.

Water coolers are placed in each dug out, and the announcer's station is set up.

Hot dogs are sizzling on the grill in concessions, and music fills the air.

As the All Star bunting is placed on the fence, players start to arrive, ready to play their favorite game.

It's game day, the best day of the week.

All Star player, Colby Maha is excited!

He is ready to hit the ball all the way to outer space.

It's game day, time to put that conditioning into play.

Caleb "The Cannon" Gallagher plays hard in his adult league, and also comes to help the younger players.

Warming up with a few knuckle balls, Caleb is ready.

It's game day, and the game needs the happiest player in the league, Destiny Otricelli.

Destiny's smile lights up the field and makes the game even more fun.

It's game day.

Special Olympics swim star,
Olivia "Pizza" Macurak,
is ready to hit a home run for her
All Star team.

It's game day.

The players are ready.

The coaches are ready.

The buddies are ready.

It is almost time to start.

The stands are full of fans.

But where is the announcer?
The All Star announcer is never late.

It's game day.

The Miracle League All Stars head to their dug out.

It is a crisp cool afternoon, perfect weather for baseball.

Caleb thinks about the crack of the home run ball off the bat. He tells his team, "Baseball is the best sport in the entire planet."

It's game day.

Colby is ready to swing hard and run all the way home.

He knows he can do it with help from his teammates.

It's game day.

Olivia thinks she will hit her coach with the ball. It happens every game, but she knows it won't hurt.

Olivia likes playing outfield as long as she doesn't have to wear a glove. She has a great bare-handed catch.

It's game day.

Destiny watches the hustle and bustle as the field is being set up for play. Players check in with their coaches and talk with their friends.

Destiny isn't sure about hitting the ball, but knows she can do it. Rounding the bases after a hit is the most fun of all.

It's game day.

But why was the white board empty? Why was the microphone silent?

It was game time. Where was the announcer?

"I'm coming!" The players turn when they hear that familiar voice.

The All Star announcer, Garrett "G" Willner comes running to take his spot. Everyone knows he would never miss a game. Garrett fills in the white board with the line-up. It is time to start.

It's game day.

Taking the microphone, Garrett sets the tone for a great afternoon.

"Good afternoon, everyone. This is Garrett Allan Gregory Willner announcing today's game. Welcome to the All Star game between the visiting team and the Miracle League All Stars at beautiful Casey's Clubhouse."

As the players take the field, everyone knows it is time to Play Ball!

It's game day at the Miracle League of the South Hills.

The dugouts and stands fill with cheering fans and encouraging words.

The field fills with smiles and laughter, home runs and catches, with players, coaches, and buddies.

And most of all, the field fills with love.

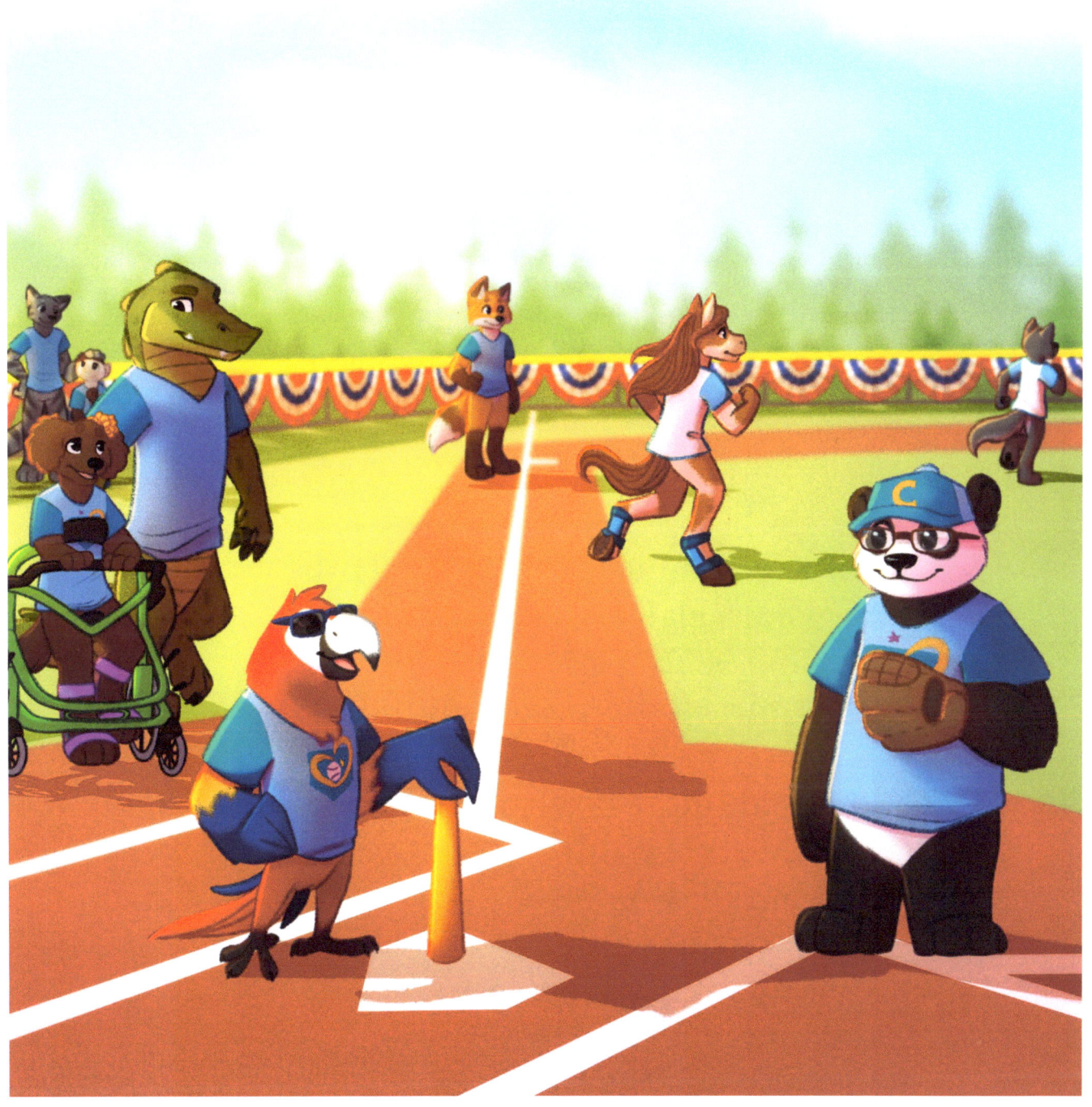

Miracle League of the South Hills
www.miracleleaguesouthhills.org
412-660-9976
info@miracleleaguesouthhills.org

Physical Address of the Field and Playground:
1551 Mayview Rd
Pittsburgh, PA 15241

Mailing Address:
1023 Paxton Dr
Bethel Park, PA 15102

# Meet our All Stars

STRIKE OUT H E
HOME
INNING

CCM

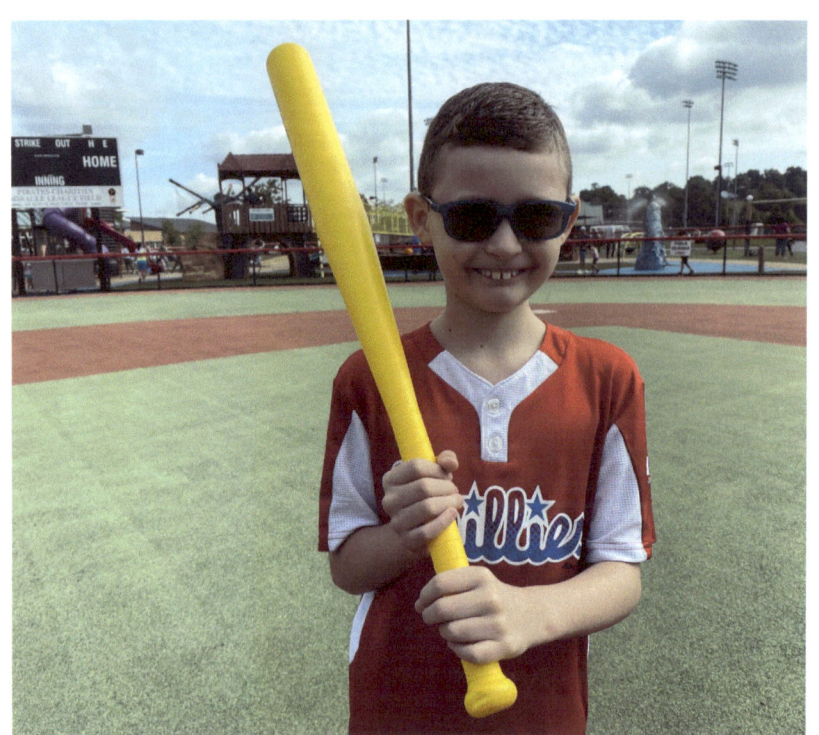

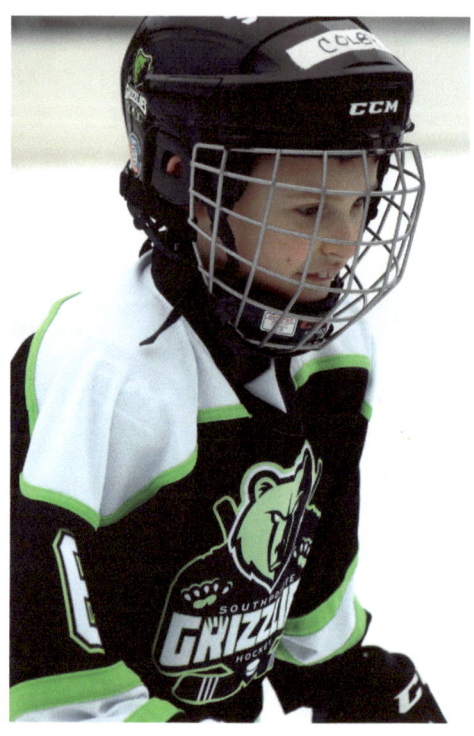

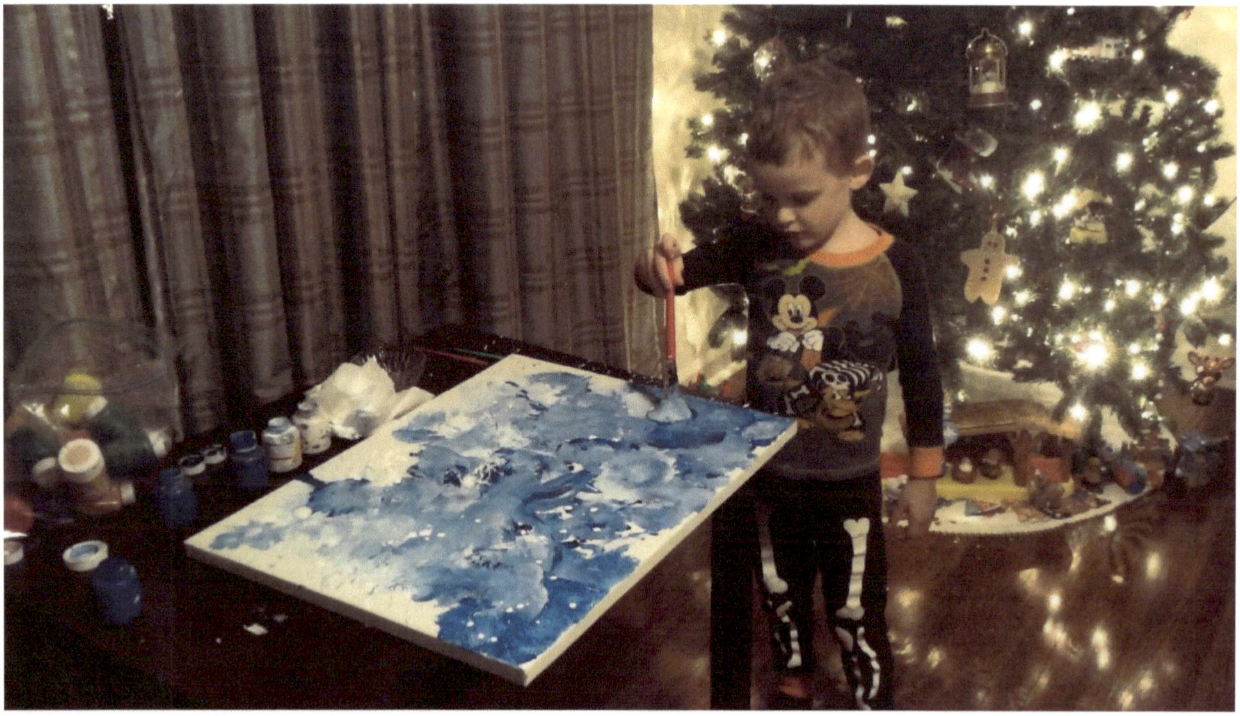

**Colby Maha** is a 10 year old boy with Autism, born and raised in the suburbs of Pittsburgh, Pennsylvania. Colby loves baseball, hockey, and painting. Colby has sold some of his paintings created as young as 3 years old. Colby enjoys spending most of his time learning about animals on his I-pad. Colby attended the Goddard School in Murrysville, PA, which was instrumental in getting Colby the resources he needed. It was also discovered by his wonderful teachers that Colby had a flair for painting! Colby currently attends the Hope Learning Center (Hope Academy) in the South Hills. Understanding language is Colby's biggest challenge. His exceptional teachers and therapists have him blossoming verbally inside and outside of school. Colby has become quite the social butterfly!

Colby also loves to attend Sunday school and VBS at Waterdam Church. He truly loves people and is most comfortable around his family, including his many cousins, aunts, uncles and grandparents. Colby is very active and loves sports. He has said on many occasions "I play two sports, baseball and hockey." He comes from a family where sports are very important. Colby is the only one in the family to take up hockey. Since people and socialization are important to Colby, sports offer the perfect outlet for his social development. Thanks to baseball at the Miracle League of the South Hills, and their fantastic coaches, Colby has become more outgoing and social, making many friends! Go Phillies! Through connections at the Miracle League of the South Hills, the sport of hockey (Hockey Sticks Together Special Hockey Program.) was introduced. Go Grizzlies! Colby is very outgoing and loves making new friends in baseball and hockey!

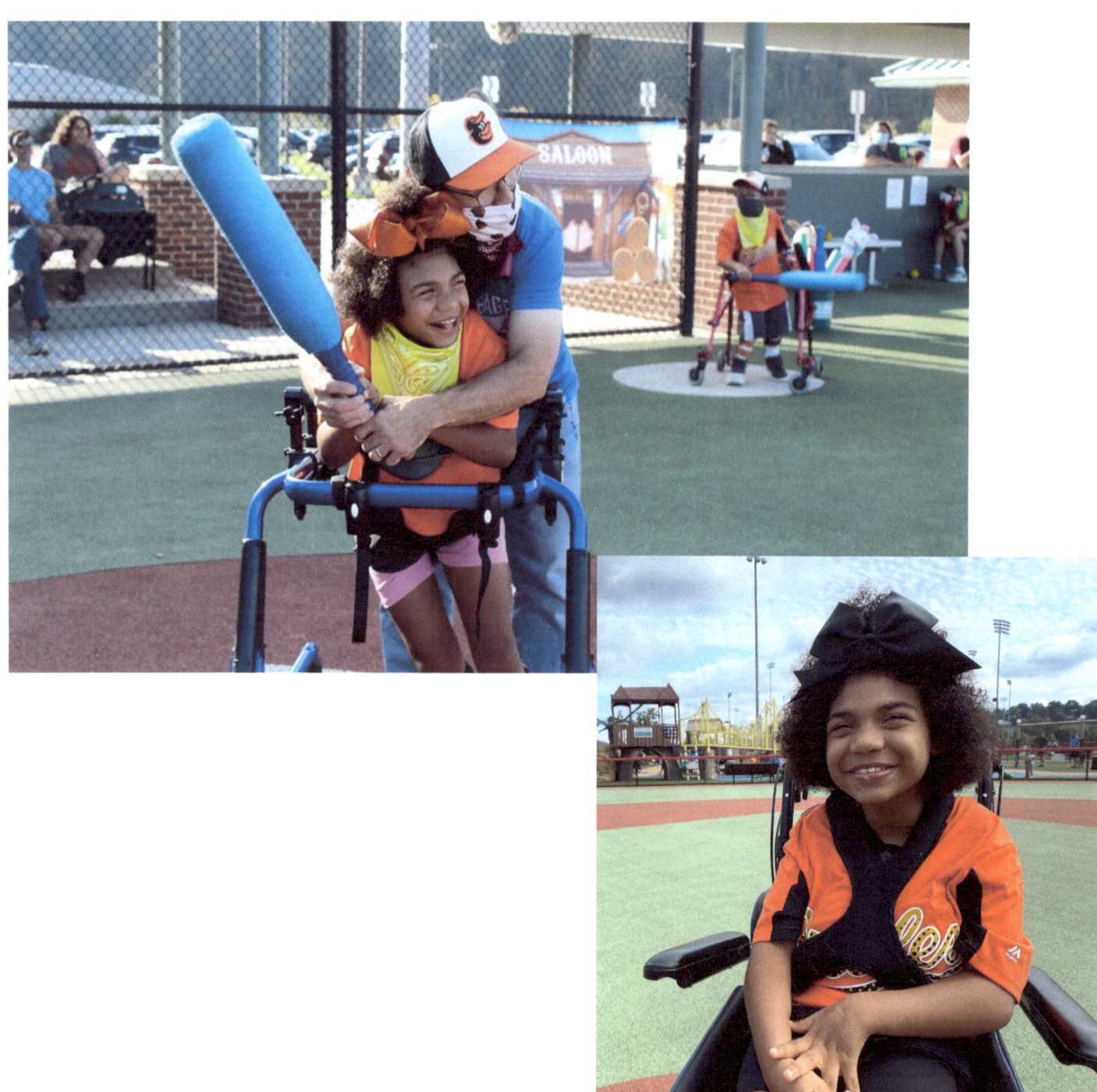

**Destiny Otricelli** is 13 and was diagnosed with cerebral palsy at the age of one due to a lack of oxygen before she was born. She has had many medical setbacks, which have limited her, but she always comes back smiling.

Destiny is nonverbal, but if you ask anyone who knows her, they'll tell you her smile says more than any words can. It is infectious; she makes friends wherever she goes.

She loves playing for the Miracle League, where she is dubbed "the happiest player in the league." She also enjoys cheering and attending school where everyone knows her name.

Destiny may have some limitations, but she has one great strength, and that is that she can touch so many with her pure unadulterated joy. No matter the physical and medical challenges she will face, it never takes away her happiness.

Hi, I'm **Caleb Gallagher**. My grandma, who raised me, loves to tell the story of how little I was when I was born. I fit right in her hand, and she said I looked like a baby bird without any feathers. Grandma and Pop-Pop spent lots and lots of time making sure I grew up okay. Growing up we had a cabin in the woods where I would spend lots of time riding my bike around and around the cabin. It took me a long time to learn how to ride my bike. My Pop-Pop stayed with me until I could fly off into the wind.

My brother Alex and my sister Lillie were my best friends. With Alex by my side, I was able to play tag football. Once I grew big and strong, Alex would have me protect him from the other guys trying to get the ball. It was hard for me and Grandma when Pop-Pop passed on, but I always helped her out when things were too heavy for her to carry.

I loved going to my school. In my senior year, I was elected Mayor of the school. All the kids there were great and today me and Elijah are still best friends. I attended the community college and graduated with a culinary certificate.

Growing up, I learned that even if things are hard, you just keep trying and everything will be okay.

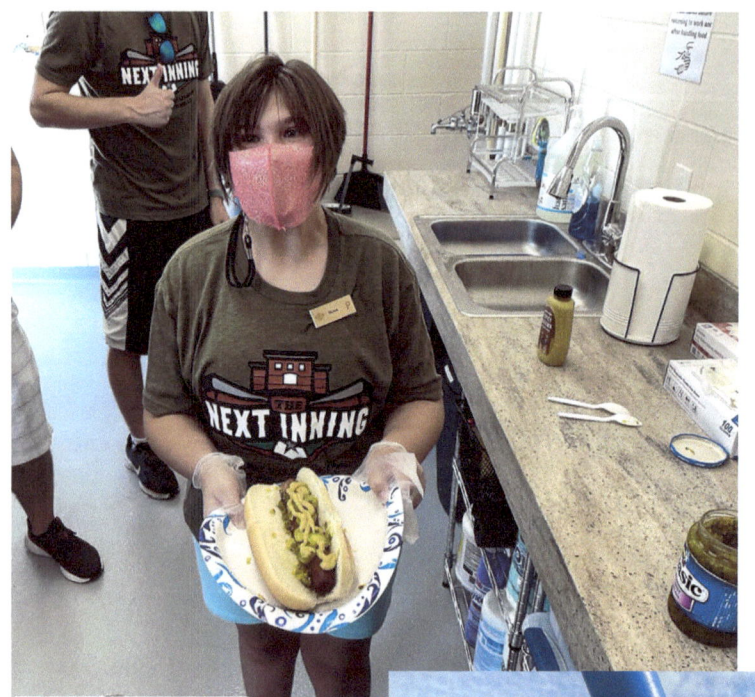

**Olivia Macurak** was born and raised in Pittsburgh, PA, and is a senior at Carrick High School. Olivia is an avid sports fan and cheers for her teams: the Penguins, the Pirates, and especially the Steelers. JuJu is her favorite player, and she was able to meet him and run out onto the field waving her Terrible Towel.

Olivia really enjoys playing baseball at the Miracle League of the South Hills. She is also an avid swimmer and has won numerous awards in Special Olympics. She really enjoys being with her friends. Olivia was excited to start her first job at the Next Inning at the Miracle League.

Olivia has a good heart and a soft soul. She has a motherly instinct and likes to help people and make sure they are okay. She wants to work in a doctor's office someday because she wants to help kids feel safe and let them know everything is going to be okay.

**Garrett Willner** loves to play baseball, and, unlike his character in the story, he is always early to games because he loves to help his coaches write the lineup on the white board before his teammates arrive.

He enjoys playing catcher and pitcher for his Tigers team, but has become known as the announcer for the games, welcoming everyone to the field. His favorite part of the game is swinging the bat.

Besides baseball, Garrett enjoys swimming and riding horses. He has a dark brown horse named Mickey that he helps tend.

# About the Illustrator

**Chelsea Rae Meeker** is a 24-year-old illustrator from Madrid, Iowa. Chelsea was diagnosed with Central Auditory Processing Disorder (CAPD), renamed Autism Spectrum Disorder (ASD), at age five. Her homeschool journey began after being told she could not learn in a classroom at the age of six.

At the age of 13, Chelsea discovered her love for art and began to dream that someday she would write and illustrate children's books. Shortly after she was diagnosed with Hypercacusis, which made everything painfully loud. She found her voice and comfort in her art. After years of struggling with an invisible disability, a passion grew within her. She began using her special abilities to help others who also struggle with feeling invisible.

If you would like to see more of Chelsea Rae's art, check out her Instagram page @chelsea_rae_arts.

# Our Vision, Mission, and Core Values

### Vision
The Miracle League of the South Hills envisions a community where all people with disabilities have opportunities to enhance their lives as athletes, teammates, and in everyday life.

### Mission
The Miracle League of the South Hills creates an environment of acceptance for people with disabilities and their families, with baseball as the root, as we promote meaningful, authentic relationships that connect people on the field as well as in the community.

### Core Values
Provide every child and adult the opportunity to play the game of baseball

Nurture peer relationships as we merge communities together so that barriers to friendship and opportunity are removed while promoting teamwork, inclusion, and helping people reach their full potential

Develop life skills that empower and challenge everyone to foster acceptance of all individuals

www.ingramcontent.com/pod-product-compliance
Lightning Source LLC
Chambersburg PA
CBHW041543040426
42446CB00003B/219